Mériam Sabbah
Dalila Gargouri

Shared medical records in the era of computerized medical records

Mériam Sabbah
Dalila Gargouri

Shared medical records in the era of computerized medical records

ScienciaScripts

Imprint
Any brand names and product names mentioned in this book are subject to trademark, brand or patent protection and are trademarks or registered trademarks of their respective holders. The use of brand names, product names, common names, trade names, product descriptions etc. even without a particular marking in this work is in no way to be construed to mean that such names may be regarded as unrestricted in respect of trademark and brand protection legislation and could thus be used by anyone.

Cover image: www.ingimage.com

This book is a translation from the original published under ISBN 978-620-6-71812-3.

Publisher:
Sciencia Scripts
is a trademark of
Dodo Books Indian Ocean Ltd. and OmniScriptum S.R.L publishing group

120 High Road, East Finchley, London, N2 9ED, United Kingdom
Str. Armeneasca 28/1, office 1, Chisinau MD-2012, Republic of Moldova, Europe
Printed at: see last page
ISBN: 978-620-8-08695-4

Contents

Introduction

The main aim of the Shared Medical Record (DMP) is to provide healthcare professionals, with the patient's prior consent, with medical information (medical history, biological test results, imaging, current treatments) from other healthcare professionals (family doctors, specialists, nurses or hospital staff) defining a medical profile for each patient [1].

The purpose of the DMP is to provide the attending physician with the most complete information, so that he or she can propose the most appropriate treatment or examinations, and also to avoid unnecessary duplication of complementary examinations or prescriptions. Another benefit of its use is that it enables health data to be shared securely, with the aim of better coordinating care [2,3]. In the future, it could also be of epidemiological interest for the early detection and prevention of health problems.

Digitization in healthcare is a national issue in Tunisia [4]. The Ministry of Public Health has made the development of digital health a strategic focus of its reform plan. The digital health development program involves the implementation of a hospital information system (setting up computerized medical records (DMI) in facilities), as well as the development of telemedicine (teleconsultation and (Appendix 1).

At Habib Thameur Hospital (HHT), the DMI was introduced in 2015; initially without data sharing between the different departments. More recently, in 2019, the medical record was shared between the different hospital departments (figure 1). This sharing was carried out following approval by the medical committee. It concerns the list of patient admissions and paraclinical explorations (figures 2 and 3).

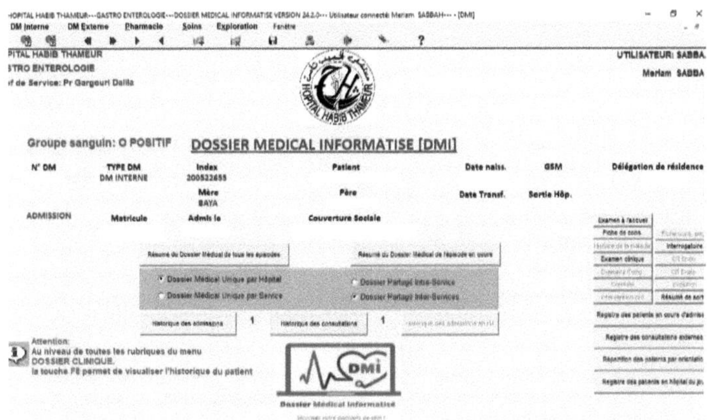

Figure 1: Sharing data in the HHT DMI: single medical record per hospital and shared record between departments.

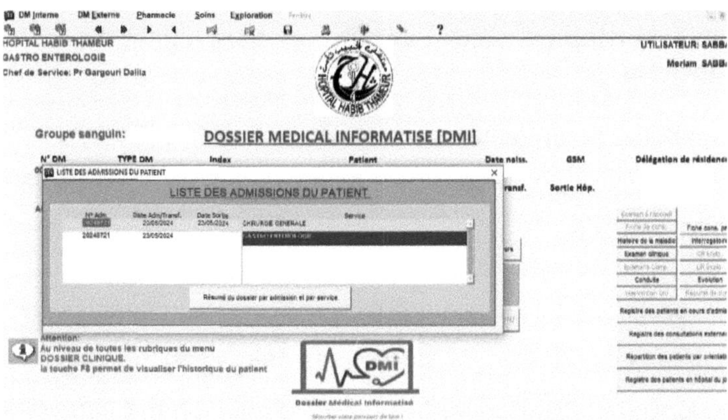

Figure 2: Interdepartmental sharing of hospital admissions (example of a patient hospitalized in different departments)

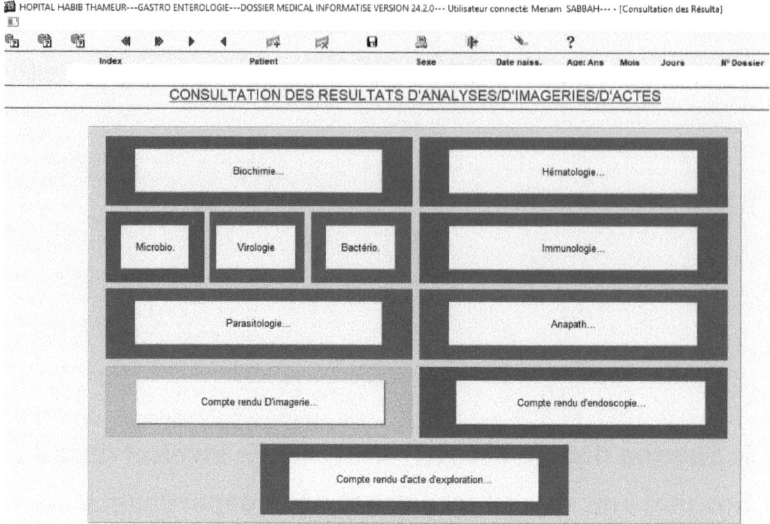

Figure 3: Interdepartmental sharing of para-clinical investigations

From an ethical point of view, the DMP is a tool of collective and, a priori, individual utility, based on the principles of beneficence and respect for autonomy.

The main obstacles to its use are its cumbersome nature, unstructured content, the accumulation of documents that cannot be modified, its non-ergonomic nature, leading to non-adherence on the part of doctors, and the lack of information provided to patients about its use [1,3].

This lack of information raises ethical questions about its use. Indeed, respect for medical confidentiality, patient consent for its implementation and the sharing of health data, as well as trust - the basis of the relationship between caregivers and patients - are all values put at stake by the DMP.

These ethical aspects have received little attention in the literature, especially in Tunisia. Hence the interest of our study.

The aim of our work was to describe, through the experience of the HHT, the ethical issues raised by the sharing of health data, in particular information and confidentiality, and to identify the type of data and the objectives of data sharing (care, research, etc.).

Methods

1. Type, location and period of study :

We conducted a descriptive prospective cross-sectional study at HHT over a seven-month period (August 2023 - February 2024).

2. Study population :

2.1 Inclusion criteria: we included all graduate doctors and doctors in training using the DMI at the HHT who had given their explicit informed consent to participate in the study.

2.2 Exclusion criteria: incomplete questionnaires were excluded from the study.

2.3 Non-inclusion criteria: questionnaires distributed but not returned after distribution were not included in the study.

3. Methodology :

The Information Optimization Commission is a HHT committee whose aim is to improve DMI management. Its members are the referring physicians in each department. A pre-established questionnaire (Appendix 2) was sent to study participants either by email, or delivered by hand via the referring physicians on the committee, or door-to-door in the various departments.

The questionnaire included :

- General information (gender, year of birth, grade, years of experience, computer software skills)

- Information about the DMI (applications and headings used, impact on daily practice, positive and negative points of health data digitization)

- Information about the DMP (healthcare personnel and data concerned by sharing)

- From information on the ethical issues raised by the DMP (ethical dilemmas raised by the sharing of health data)

- Finally, an open-ended question was asked at the end of the questionnaire to gather suggestions from participants.

4. Data analysis and statistical study :

Data entry and statistical analysis were carried out using Statistical Package of the Social Sciences for Windows version 25. A descriptive study was carried out, with calculation of simple and relative frequencies (percentages) for qualitative variables, and means and range (extreme values: minimum and maximum) for quantitative variables.

5. Bibliographic research :

A bibliographical search was carried out on the Pubmed and Science Direct sites, using the following keywords: computerized medical record, shared medical record, ethics. We also searched the library of theses and dissertations of the Tunis Faculty of Medicine for works on the same subject.

The references were managed using ZOTERO® bibliographic reference software.

6. Ethical considerations :

Throughout the study, the anonymity of the healthcare professionals included was respected.

Written informed consent in French was obtained, and completed by the participants in the first part of the questionnaire provided (Appendix 2). The agreement of the HHT institutional ethics committee was obtained for the study (Appendix 3).

We declare no interest in this work.

Results

1. Study population

Of the 100 questionnaires handed in or sent by email, we received 86 responses, for a participation rate of 86%. No participant was excluded from the study (Figure 4).

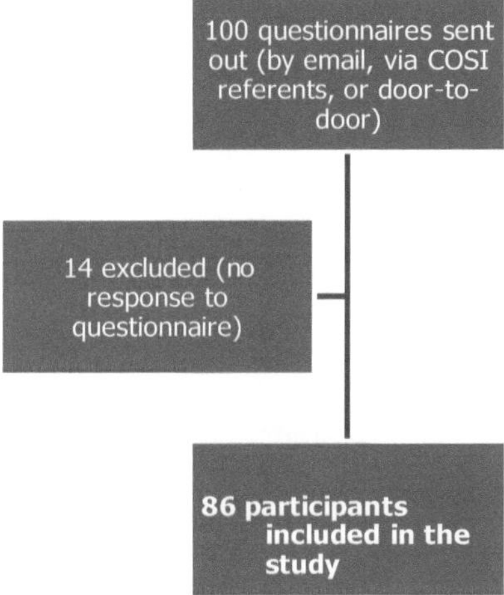

Figure 4: Study inclusion diagram

2. General population characteristics

The 86 participants were divided into 35 men and 51 women, with an M/F sex ratio of 0.43 and a mean age of 33.5 years [25-56 years]. The general characteristics of the study participants are summarized in the table below (Table I).

Table I: General characteristics of study participants

Features	Data	Number (n)
Service	Anesthesia and intensive care	8
	Anatomopathology Cardiology	1
	General surgery Pediatric	20
	surgery Gastroenterology	11
	Ophthalmology	14
		27
		2
	Internal medicine	3
Professional grade	Professor	7
	Associate Professor Assistant	13
	university hospital	17
	Specialist Resident	5
	Internal	29
		15
Mastery previous of computer software	Yes	86
	No	0
Type of software mastered	Word Excel PowerPoint	86
	Statistical software (SPSS)	60
		74
		49

3. Knowledge of DMI

Participants' responses regarding the use of DMI in their daily practice are summarized in the table below (Table II).

Table II: Data on DMI use

Question	Answer	Number (n)
Practical use of the DMI daily	Yes No	85 1
Application used	DMI DME Access appointment management Application of anatomopathology	85 15 10 3 1
Aims of DMI use	For patient care For data collection for scientific work	85 18
DMI bricks used	Medical observation Discharge summaries Monitoring Investigations (laboratory, radiology) Entering procedure/operative reports Inventory management Pharmacy Coding Service statistics	71 56 33 85 55 5 70 45 2

DME headings used	Enter consultation Request additional	45
	tests Medical prescription Book	86
	appointment Medical certificate	55
		12
		16
Management items	Outpatient appointments	62
appointment	Appointments for procedures (endoscopy...)	70
used	Hospitalization appointments	22
	Leave management	3
Daily practice	Yes	80
facilitated by the DMI	No	6
Positive aspects of	Improves patient follow-up Facilitates	82
digitizing healthcare	multidisciplinary management	80
data	Rationalize complementary examinations	76
	Enables medication tracking Facilitates	59
	retrieval of results of additional tests	78
	- Imaging	
	- Endoscopy	
	- Biology	
	- Anatomopathology	
Negative points of	Lack of equipment (computers) Lack of	37
using digitizing	user training Time-consuming	42
applications	Computer bugs	21
health data	Patient identification	28
		10

4. Knowledge of the DMP

Participants' responses regarding their knowledge of the DMP are summarized in the table below (Table III).

Table III: DMP data

Question	Answer	Number(n)
Nature of DMI sharing	Single DM per hospital Single DM per department Intra-departmental shared file Inter-departmental shared file I don't know	44 21 11 38 5
Data concerned by sharing	Observation summary Discharge summary Outpatient form Additional tests Surgical report Preoperative consultation anesthetic	67 71 22 78 33 2
Healthcare personnel to be concerned by sharing	Between doctors Administrative department Registration department Financial department Nurses Nutritionists With the technicians anesthesia	86 44 35 21 69 12 17
Improvements related to health data sharing	Comprehensive patient care Multidisciplinarity in patient care Resource management Archiving	68 79 80 55

5. Ethical dilemmas raised by the DMP

The final part of the questionnaire dealt with the ethical issues raised by the DMP. The ethical reservations expressed by participants concerning the DMP were mainly linked to the values of :
- confidentiality and respect for medical secrecy (n=70; 81%),
- communication / information (n=48; 56%)
- ethical publications (n=11; 24%).
- No ethical reservations were noted among five participants (6%).

In addition, 60 participants felt that informed consent was required before sharing health data (69%). These results are summarized in figure 5.

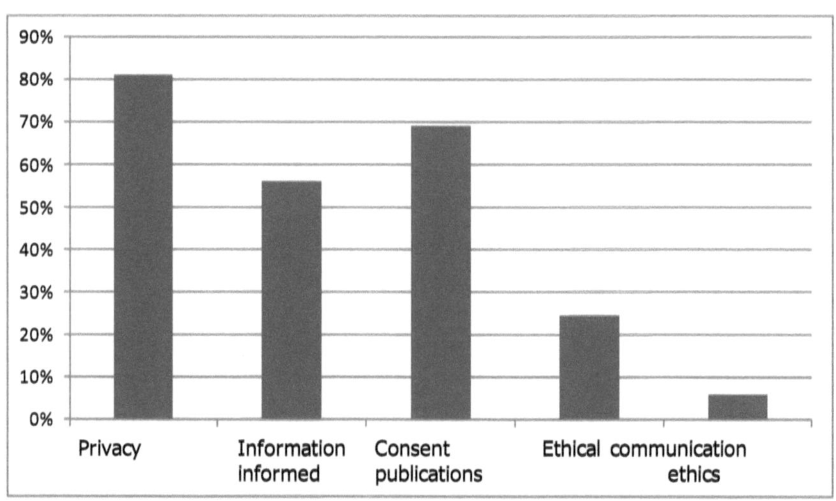

Figure 5: Participants' ethical reservations about the DMP

6. Suggestions made by participants

At the end of the questionnaire, participants were invited to write their own suggestions. The main suggestions were :
- Proposed training courses in the use of DMI for new interns/residents and paramedical staff
- Insertion of an attachment section to integrate explorations carried out by patients in the private sector and liaison letters from referring physicians
- Improved coding list (transition to the eleventh version of the International Classification of Diseases)
- Application customization by specialty
- Validation of pre-established forms to facilitate computerization o f data, considered time-consuming
- Total inter-hospital and inter-departmental sharing of healthcare data to improve multidisciplinary patient management
- Availability of tablets to facilitate handling and use of the DMI during bedside visits
- Improved computer network (faster connections, fewer bugs, easier access to X-ray images)
- Training of registration staff on the importance of respecting the unique patient identifier to avoid duplicate records for the same patient.
- Proposal to introduce an informed consent form for patients concerning the sharing of their health data.

Discussion

In this chapter, we will review the main findings of our study, then list its strengths and limitations. We will then define the DMP, specify its legislative framework, before raising the ethical dilemmas raised by its use and ending by proposing recommendations and perspectives in current clinical practice linked to the DMP.

1. Key results

We conducted a descriptive cross-sectional prospective data collection study at the HHT over a seven-month period (August 2023 - February 2024), the aim of which was to describe, through the HHT's experience, the ethical issues raised by health data sharing, in particular information, confidentiality, and identification of the type of data and the purposes of data sharing (care, research...).

We included all healthcare professionals (qualified doctors and doctors in training) using the DMI at the HHT who had given their written informed consent to participate in the study. Incomplete questionnaires were excluded from the study, and questionnaires distributed but not returned after distribution were not included in the study.

The questionnaire was sent by e-mail and delivered door-to-door to doctors with a degree or in training in the various departments of the HHT. It included general information, information on the DMI, information on the DMP and information on ethical issues raised by the DMP, as well as a question open to suggestions.

Of the 100 questionnaires sent out or handed in, we included 86 participants, i.e. a participation rate of 86%, divided into 35 men and 51 women with a M/F sex ratio of 0.43 and an average age of 33.5 years [25-56 years].

Specialties were dominated by gastroenterology (n=27), followed by cardiology and pediatric and general surgery (n=20, 14 and 11 respectively). Residents were in the majority in 29 cases. Previous mastery of computer software was noted in all participants, with a preference for Office followed by statistical software. Use of the DMI was common practice in 85 cases, with the DMI as the main application, followed by the EMR and appointment management. The use of the DMI facilitated routine practice for 80 participants, the negative points being mainly the lack of resources (computers), training in the use of the DMI, and computer bugs in 42, 37 and 28 cases respectively.

Forty-four of the participants thought that the medical record was unique per hospital, while 38 thought that data sharing was inter-departmental. According to the participants, data sharing should mainly concern explorations (n=78), discharge summaries (n=71) and observation records (n=67). They also all agreed that data should be shared between doctors. Furthermore, sharing health data optimized resource management, multidisciplinarity and overall patient care in respectively

The ethical reservations expressed by participants concerning the DMP were mainly linked to the values of confidentiality and respect for medical secrecy (n=70; 81%), followed by communication/information (n=48; 56%). Sixty participants felt that informed consent was required before health data could be shared (69%).

2. Study strengths and limitations

2.1 Study strengths

Our study represents the first Tunisian study to focus on the DMP and specifically on the ethical considerations it raises. Another strength of our study is that it is cross-sectional and prospective. In addition, it included a homogeneous population of participants from the medical field. In addition, it focused on an innovative subject, testifying not only to the pilot nature of HHT's efforts to digitize health data, but also to the structure's interest in ethical considerations related to care.

2.2 Study limits

Our study nevertheless has a number of limitations. Indeed, the small sample size and monocentric nature of the study reduce the interpretability of the data, especially as the DMI is currently used in the majority of hospital structures. Similarly, a study encompassing the various healthcare professionals (paramedical, administrative) involved in patient care and who use the DMI routinely in their daily practice would have been interesting to conduct.

3. General information on the DMP

3.1 Definition

The DMP is a secure storage space for health data, enabling healthcare professionals to share documents relating to patients' health. Healthcare professionals and establishments can enter and consult their patients' DMPs [3].

3.2 Shared data

All data held by healthcare professionals can be shared, provided they are useful for patient care. This may include, for example, results of explorations (biological, radiological); medical prescriptions, hospitalization or outpatient reports, operative or exploration reports, or vaccination or medication schedules. In all cases, sharing should be limited to "necessary, relevant and non-excessive" data directly related to each healthcare professional's area of expertise, each of whom is bound by strict medical confidentiality [2].

3.3 Persons with access to data sharing

This sharing concerns healthcare professionals with whom the patient has a care or therapeutic relationship, i.e. within a strict framework of quality and continuity of care. Thus, shared secrecy is a new concept highlighting the various parties involved in the relationship between caregiver and patient, and no longer concerns doctors alone, but involves all healthcare professionals (paramedical teams, cleaning staff, ambulance drivers, medical secretaries, etc.).

...). Multidisciplinary practice has further accentuated this shared secrecy. [1,3]. There is, however, the question of sharing with certain people, such as the administrative or financial department, the supervisory authority (Ministry of Public Health) or insurance companies.

4. DMP legislative framework

Tunisian data on the subject are fairly scarce. As part of the XVIIIth summit of the Francophonie held in Djerba in 2022, one of the objectives of the Ministry of Public Health, as part of the program to develop digital health in Tunisia, was the deployment of the Electronic Medical Record (DMI) in public health structures to improve patient care (better follow-up thanks to a patient history, continuous real-time sharing of medical data between the various parties involved in the patient's care, etc.)[4].

Previously, in a study published in 2018, concerning the acceptability of the DMI by doctors in hospitals [5], the results suggested that decision-makers in Tunisian hospitals needed to guarantee a working context favorable to optimizing the use of IT by informing and explaining the benefits and opportunities, and by fostering a professional environment supporting the sharing of information respecting the legislative and ethical framework of the medical sector.

The French data protection authority also issued an advisory opinion on the DMP in its 2022 toolkit [6], stressing the importance of respecting personal data by anonymizing it as far as possible, based on the 2004 law on the protection of personal data [7]. It also stressed the need for a high level of authentication to enable access to healthcare data shared between different professionals (single-use password or any other two-factor authentication mechanism (smart card, USB key, etc.) to guarantee IT security. In addition, this body insists on the need to obtain an agreement for the international sharing of medical data.

The Tunisian Code of Ethics [8] does not include a specific paragraph on the DMP, but its eighth article stresses the confidentiality of doctors, specifying that all doctors are bound by professional secrecy, unless otherwise stipulated by law. This reflects the absence of a legal or regulatory framework for shared confidentiality, as opposed to the general and absolute nature of medical confidentiality.

Elsewhere, as in France for example, the legislative framework is more precise, set out in Decree no. 2021-1047 of August 4, 2021 on the shared medical record [9]. This specifies in three sections the creation and content of the DMP, the terms of access and rights of the holder, and finally the rights of authorized professionals. Back in 2006, the French Data Protection Authority (Commission nationale de l'informatique et des libertés) issued a decree [10] on the hosting of personal health data. Marcelli et al proposed a number of solutions for securing the DMP [11]. They proposed that each patient should become his or her own host, via a mobile device, memory card or key, which would constitute his or her file. At each consultation, the patient would provide his or her carte vitale and file, and enter a personal password. The practitioner, for his part, will need to have the software to operate the medium and access it on his computer using his professional health card. The encrypted medical information that appears can only be used by the attending practitioner and the patient, since it requires simultaneity of place and time between the medium, the carte vitale, the patient's secret code and the presence of the doctor with software and security codes.

In Belgium, too, the DMP is governed by a clear law: the law of April 22, 2019 on the quality of healthcare practice [12]. This specifies that only healthcare professionals who have a therapeutic relationship with patients have access to their personal health data. In addition, this law encourages compliance with the following conditions: the purpose of access is to provide healthcare; access is necessary for the continuity and quality of the healthcare provided; access is limited to data that is useful and relevant to the provision of healthcare.

5. Ethical considerations of the DMP

The sharing of healthcare data raises questions about **confidentiality and respect for medical secrecy.** Indeed, when a patient entrusts his or her health data to a healthcare professional, the data is shared with an entire healthcare team, including medical, paramedical and administrative staff [13].

Here, too, the patient is subject to an internal **conflict of** conflicting **interests**, between keeping the secret and the information to himself in order to preserve his privacy, or disclosing it to obtain the best care [11].

Furthermore, the computerization of my data poses the problem of cyber-security, ensured by highly secure access and storage on the DMI network. For example, secure access using a password or a professional card reduces the risk of piracy [14].

Respect for confidentiality and medical secrecy requires **informed consent** to data sharing. This is the patient's agreement to allow his or her health data to be shared electronically and securely between the people involved in his or her care. The sharing of such data takes place exclusively within the framework of continuity and quality of medical care, and in compliance with regulations governing the protection of personal data [1,2,15].

Consent for the processing and sharing of health data should not be confused with the consent required for the performance of certain medical acts [16,17]. It must be informed, explicit and express. An example of consent for health data sharing created by the European Network for Rare Diseases is given in Appendix 4 [18]. Such sharing would help to improve patient care by, for example, identifying cases of an orphan disease.

Sharing information is therefore an appropriate means of providing quality care, but it can only be done with the patient's informed consent. This implies providing the patient with all useful details concerning the aims of sharing, the envisaged content, the function or remit of the recipient institutions, and even the names of the interlocutors, as well as assessing the stakes and possible consequences for the patient's situation of sharing or not sharing certain information [19].

Importantly, patients can withdraw their consent at any time, or refuse certain healthcare professionals access to their data. Similarly, they can ask the healthcare professional concerned not to share certain information. This is part of the broader context of respect for patient **autonomy** and the relationship of trust that patients have with healthcare professionals, which is fundamental to the relationship between healthcare professionals and patients [20].

At present, informed consent is not obtained for data sharing at the HHT; it is presumed to have been obtained, unless the patient objects. However, this raises the question of **how to inform and communicate** with the patient about data sharing and its modalities [21].

Firstly, patients must be informed about the sharing of their health data, and this information must be clear, fair and appropriate. Information is the source of consent and its validity.

On the other hand, information and communication between the various healthcare professionals via the DMP will optimize patient care within a multi-disciplinary framework, and will also benefit public health by reducing healthcare costs associated with redundant explorations (biological or radiological, for example). This optimization of health resources, through the control of health expenditure, is rooted in an overall spirit of **benevolence** towards the general population, and according to utilitarian theory, privileging the good of the majority over personal interest [17]. The aim of sharing is to improve the efficiency and quality of care.

In addition to the ethical values raised, it is important to note the problem of the **ethics of publications** that may be implicated by the DMP [22]. Indeed, certain health data could be used for scientific purposes, in the ignorance of patients and without the information and involvement of certain healthcare professionals who have participated in patient care. This could be mitigated by training healthcare professionals in publication ethics.

6. Outlook and recommendations

At the end of our study, we can propose a few recommendations concerning the use of the DMI, the DMP and the ethical considerations that these applications raise.

In fact, we propose training sessions for healthcare professionals on the ethical values raised by the sharing of health data. These training sessions could be carried out under the aegis of the hospital's continuing professional development committee, in association with the institutional ethics committee. They would raise awareness among all healthcare professionals (medical, paramedical and administrative) of values such as confidentiality, communication and publication ethics.

Furthermore, the promotion and development of the DMP can only be envisaged through the development of standards and protocols common to all health establishments, and the development of IT systems with secure storage and the implementation of strict confidentiality and security rules to protect patients' personal and medical information. The same applies to the hosting of health data, which must be highly secure.

We therefore propose that the DMP should be limited to certain information required for multi-disciplinary collaboration and continuity of patient care, common to all hospital structures. This information includes

- Antecedents
- Hospitalization reports
- Care history for the past 24 months
- Results of investigations (biological and radiological analyses, procedures, operative reports, etc.)
- Emergency contact details
- And possibly, advance end-of-life directives if these have been defined by the patient...

In addition, as suggested by some participants, it would be interesting to offer patients an informed consent form, in which they would give their authorization for their medical data to be shared between the various professionals involved in their care. We therefore propose a consent form to be inserted in the DMI and completed by the patient (or his/her legal guardian) as soon as he/she is consulted or hospitalized at the HHT, with an electronic signature (appendix 5).

Finally, the prospect of an inter-hospital DMP (or even inter-healthcare structures, thus including front-line basic health centers and private-sector establishments) should be considered by the health authorities, in order to facilitate access to care for patients across the country's different hospital structures, while optimizing healthcare expenditure. Such sharing must, however, benefit from high-security data protection to guarantee the confidentiality of health data.

Conclusions

The digitization of healthcare data is a national issue. The main aim of the DMP is to make medical information available to healthcare professionals, with the patient's prior consent, in order to optimize medical treatment and ensure continuity of care in a multidisciplinary context. This sharing of information brings ethical values into play, including confidentiality, informed consent, information and communication. These ethical aspects have received little attention in the literature, especially in Tunisia. Moreover, the DMP was recently introduced at the HHT following approval by the medical committee, hence the interest of our study.

The aim of our work was to describe, through the experience of the HHT, the ethical issues raised by the sharing of health data.

We conducted a descriptive cross-sectional prospective data collection study at the HHT over a seven-month period (August 2023 - February 2024), the aim of which was to describe, through the HHT's experience, the ethical issues raised by the sharing of health data, in particular information, confidentiality, and the identification of the type of data and the objectives of data sharing (care, research...).

We included all graduate and trainee physicians using the DMI at the HHT who had given written informed consent to participate in the study. Incomplete questionnaires were excluded from the study, and questionnaires distributed but not returned after distribution were not included in the study.

The questionnaire was sent by e-mail or delivered door-to-door to healthcare professionals in the various departments of the HHT. It included general information, information about the DMI, information about the DMP and information about the ethical issues raised by the DMP, as well as a question open to suggestions.

Of the 100 questionnaires sent out or handed in, we included 86 participants, i.e. a participation rate of 86%, divided into 35 men and 51 women with a M/F sex ratio of 0.43 and an average age of 33.5 years [25-56 years].

Specialties were dominated by gastroenterology (n=27), followed by cardiology and pediatric and general surgery (n=20, 14 and 11 respectively). Residents were in the majority in 29 cases. Previous mastery of computer software was noted in all participants, with a preference for Office followed by statistical software. Use of the DMI was common practice in 85 cases, with the DMI as the main application, followed by the EMR and appointment management. The use of the DMI facilitated routine practice for 80 participants, the negative points being mainly the lack of resources (computers), training in the use of the DMI, and computer bugs in 42, 37 and 28 cases respectively.

Forty-four of the participants thought that the medical record was unique per hospital, while 38 thought that data sharing was inter-departmental. According to the participants, data sharing should mainly concern explorations (n=78), discharge summaries (n=71) and observation records (n=67). They also all agreed that data should be shared between doctors. Moreover, sharing health data optimized resource management, multidisciplinarity and overall patient care in respectively

The ethical reservations expressed by participants concerning the DMP were mainly linked to the values of confidentiality and respect for medical secrecy (n=70; 81%), followed by communication/information (n=48; 56%). Sixty participants felt that informed consent was required before health data could be shared (69%).

Our study represents the first Tunisian cross-sectional prospective data collection study to have focused on an innovative subject, thus testifying to the pilot nature of the HHT's digitization of health data, as well as to the structure's interest in ethical considerations related to care. However, the small sample, the monocentric nature of the study and the fact that participants were limited to doctors were all weak points.

Finally, at the end of our study, the participating doctors seem to be aware of the ethical dilemmas raised by the sharing of health data. We propose that awareness training be extended to other healthcare professionals, so as to integrate data sharing into the relationship between caregivers and patients. We also propose to collegially define the information to be shared in order to optimize the quality of care and resource management, and to insert in the DMI an informed consent form to be electronically signed by patients consulting or hospitalized in the hospital. This will enhance the value of multidisciplinarity and continuity of care through shared confidentiality. Finally, it is through awareness-raising and training of healthcare professionals and patients that the sharing of health data can be made more widespread across all hospital structures, and even between hospitals. The prospects for this study are to extend health data sharing to all caregivers, and to share health data on a national scale.

References

1. Lucas J. The sharing of personal health data in digital health uses put to the test of the person's express consent. Ethics Med Public Health. 2017 Jan;3(1):10-8

2. Hervé C, Stanton-Jean M, Martinent É. Les systèmes informatisés complexes en santé: banque de données, télémédecine, normes et enjeux politiques. Paris: Dalloz; 2013. (Thèmes & commentaires).

3. Shared medical record [Internet]. Wikipedia; 2023 Nov 17. Available from:
https://fr.wikipedia.org/w/index.php?title=Dossier_m%C3%A9dical_parta
g
%C3%A9&oldid=209639346

4. Tunisian Ministry of Health. Program for the development of "digital health" in Tunisia [Internet]. 2023 Nov 15 [accessed 2023 Nov 15]. Available from: http://www.santetunisie.rns.tn/fr/prestations/programme-de- développement-de-la-"santé-numérique"-en-tunisie?start=3

5. Hammouda SB, Hadoussa S. Projet e-santé Tunisie: étude des facteurs d'acceptation du Dossier Médical Informatisé (DMI) par les médecins auprès des hôpitaux. Management & Avenir. 2018;(102):15-31.

6. National Institute for Personal Data Protection (INPDP). Resources [Internet]. 2023 [consulted on 2023-11-28]. Available at: https://www.inpdp.tn/Ressources.html

7. Ministry of Justice. Loi organique n° 2004-63 du 27 juillet 2004

portant sur la protection des données à caractère personnel [Internet]. [cited 2023-10-26]. Available from: https://legislation-securite.tn/latest-laws/loi- organique-n-2004-63-du-27-juillet-2004-portant-sur-la-protection-des- données-a-caractere-personnel/

8. Imprimerie Officielle de la République Tunisienne (IORT). Code of Medical Ethics (CDM). Tunis : IORT; 2022.

9. Legifrance. Décret n° 2022-703 du 25 avril 2022 relatif à la mise en œuvre du Dossier Médical Partagé [Internet]. 2022. Available at: https://www.legifrance.gouv.fr/jorf/id/JORFTEXT000043914236

10. Legifrance. Loi n° 2004-810 du 13 août 2004 relative à l'assurance maladie [Internet]. 2004. Available at:

https://www.legifrance.gouv.fr/jorf/id/JORFTEXT000000264665

11. Marcelli A, Solaret D. Securing shared medical records. Bull Acad Natle Méd. 2010;194(4-5):767-78.

12. Service public fédéral Justice. Loi du 22 avril 2019 relative à la qualité de la pratique des soins de santé [Internet]. 2019. Available from: https://etaamb.openjustice.be/fr/loi-du-22-avril-2019_n2019041141.html

13. Pougnet R, Pougnet L. Le dossier médical partagé : pour un usage centré sur la personne ? Éthique & Santé. 2019;16(2):64-70.

14. Karsz S. Information sharing, between deontological constraints and ethical issues. La revue du Centre Michel de L'Hospital. 2020;(20).

15. Hecquet M. Le dossier médical, un outil de la relation de soins?

Éthique Santé. 2012 mars;9(1):11-3.

16. Pougnet R, Mazeaux S, Quere S, Pougnet L, Le Breton P, Dutray S. Ethical issues of virtual psychology software. Droit Santé Société. 2016;4(4):33-45.

17. Jousset D. The moment of decision in medical ethics. J Int Bioethique Int J Bioeth. 2014 June;25(2):113-35, 174.

18. Ern-rnd.eu [Internet]. FR_InformedConsent.pdf. Available at: https://www.ern-rnd.eu/wp-content/uploads/2018/11/FR_InformedConsent.pdf

19. Metiers.action-sociale.org [Internet]. Consentement éclairé médico-social. Available at: https://metiers.action-sociale.org/pratiques/consentement-eclaire-medico-social

20. Moutel G. Ethique et informatisation du Dossier Médical Personnalisé [Internet]. Espace de réflexion éthique de Normandie; 2020 Oct 19 [modified 2021 Aug 2]. Available from: https://www.espace-ethique-normandie.fr/9341/

21. Bonnaud G. Personal medical data on Google: real danger of digital health? Hegel. 2013;3(3):161-2. Éditions Association pour la revue HEGEL. ISSN 2269-0530. DOI: 10.4267/2042/51449.

22. Peternelj-Taylor C. Promoting Ethical Integrity in Publishing. J Forensic Nurs. 2013 Oct 25;9:65-7.

Appendices

Annex 1: Digital health development program of the Tunisian Ministry of Public Health 2017-2025.

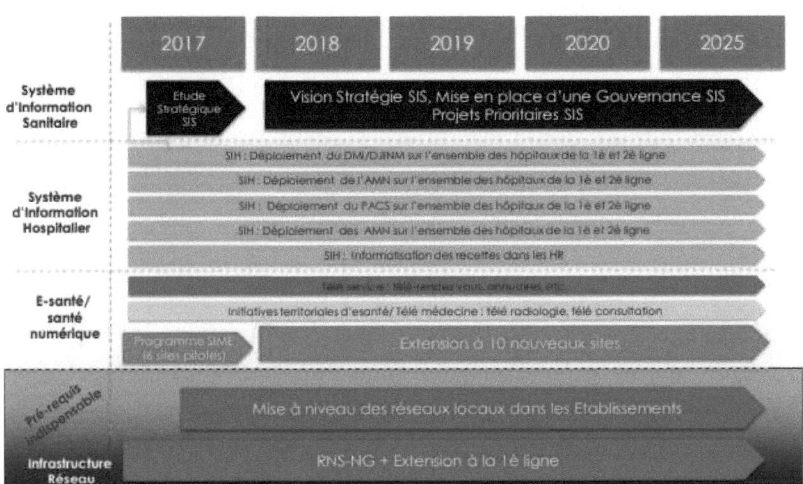

Appendix 2: Questionnaire given or e-mailed to participants with informed consent

DMI-DMP QUESTIONNAIRE

You are taking part in a cross-sectional study on shared medical records as part of a dissertation for the CEC in medical bioethics at the Tunis Faculty of Medicine, directed by Pr Ag Mériam SABBAH and supervised by Pr Dalila GARGOURI.

The approval of the HHT ethics committee was obtained for this study.

The questions you are asked will remain confidential and anonymous.

Do you agree to participate in this study?

Yes	
No	

If the previous answer is yes, please answer the questions below

General questions

Your gender	Male□ Female□
Year of birth	
You work in the	....
You work as a	Professor□ MCA□ AHU□ Resident□ Intern□ Extern □ Other□ Specify ....
Years of experience	≥ 10 years □ 5-10 years □ ≤ 5 years □
Do you know how to use computer software?	Yes □ No □ If yes, which ones?
	Word □ Excel □ Powerpoint □ Statistical software (SPSS...) □

Concerning the computerized medical file

You use computerized healthcare data in your daily practice	Yes☐ No☐
If so, which application do you use?	DMI ☐ DME ☐ Appointment management ☐ Access ☐ Other ☐ Specify …………………………………………………
You use the DMI for	Patient care ☐ Data collection for scientific work ☐
hich fields do you use in the DMI?	Medical observation ☐ Discharge summaries ☐ Monitoring ☐ Investigations (laboratory, radiology) ☐ Application ☐ Results ☐ Entering procedure/operative reports ☐ Stock management (medical devices, drugs) ☐ Pharmacy ☐ Coding ☐ Service statistics ☐ Others ☐ Specify …………………………………………………
Which headings do you use in the EMR	Enter consultation ☐ Request additional tests ☐
	Medical prescription ☐ Appointment booking ☐ Medical certificate ☐ Other ☐ Specify …………………………………………………

Which sections do you use to manage your appointments?	Outpatient appointments □ Procedure appointments (endoscopy...) □ Inpatient appointments □ Leave management □ Other □ Specify
Computerization of health has made your daily practice easier	Yes □ No □
What are the advantages of digitizing healthcare data?	Improves patient follow-up □ Facilitates multidisciplinary care □ Streamlines the performance of complementary examinations □ Enables medication monitoring □ Facilitates the retrieval of complementary examination results - Imaging □ - Endoscopy □ - Biology □ - Anatomopathology □ Other □ Specify
hat are the negative points you've encountered in using healthcare data digitization applications?	Lack of equipment (computers) □ Lack of user training □ Time-consuming □ Computer bugs □ Patient identification □ Other □ Specify ....

Concerning the shared medical file (DMP)

The DMI at HHT is a	Dossier médical unique par hôpital □ Dossier médical unique par service □ Dossier partagé intra service □ Dossier partagé interservices □ I don't know □
File sharing at HHT involves the following elements	Observation summary □ Output summary □ Outpatient consultation form □ Complementary examinations □ Operative report □ Pre-anaesthetic consultation □ Other □ Specify ....
Health data must be shared	Between doctors □ With the administrative department □ With the registration department □ With the financial department □ With nurses □ With nutritionists □ With anesthesia technicians □ Other □ Specify ....
The DMP improves	Comprehensive patient care □ Multidisciplinary patient care □ Resource management (additional tests) □ Archiving □
What ethical reservations do you have about the DMP?	Confidentiality/medical confidentiality □ Information/ Communication□ Informed consent□ Publication ethics □ Other □ Specify ....
Do you think patient consent is necessary before sharing medical records?	Yes □ No □

Do you have any comments or suggestions? If so, which ones?

....

....

....

....

....

....

....

....

....

....

Thank you for your participation in this study.

Appendix 3: Agreement of the HHT institutional ethics committee to conduct the study

Tunisian Republic
Ministry of Health
HabibThameurTeaching Hospital
EthicsCommitee

الجمهوريّة التونسية
وزارة الصحة
المستشفىالجامعيالحبيب ثامر
لجنةالأخلاقياناالطبية

HABIB THAMEUR TEACHING HOSPITAL
ETHICS COMMITTEE APPROVAL

Project Reference: HTHEC-2022-36

Project title: «Shared medical record in the era of the computerized medical record»

« Dossier médicalpartagé à l'ère du dossier médicalinformatisé»

Nature of Project: Prospective study

Lead Principal Investigator: Assoc. Prof Meriam SABBAH

Local Chief Investigator: Prof DalilaGargouri

Supervisor: ProfDalilaGargouri

Project Contact Point: Principal Investigator: phone:00 216 98629843

Mail: sabbah_meriam@yahoo.fr

Hospital / institution: HabibThameur Hospital

Department: Gastro-enterology department

Members of HabibThameur Hospital Ethics Committee (HTHEC):

Prof Ehsen BEN BRAHIM (MD, Chairperson), Prof Nebiha FALFOUL (MD, Vice-chairperson), Assoc. ProfZohraAYDI (MD, Secretary General), Prof Anis BENZARTI (MD), Assoc. ProfMeriam SABBAH, Assoc. ProfYosra ZGUEB (MD), Assoc. ProfChaouki MBARKI (MD), DrWided BEN AYOUB (MD), Assoc. ProfFatma BEN MBARKA (Pharmacist), DrHelaMAAMOURI (MD), Dr Aida DAIB (MD), Mr Adel BEN HASSINE (Lawyer), MrBilel MOUROU (Health Personnel Representative).

Date: 28/11/2022

Prof Ehsen BEN BRAHIM
Chairperson

Présidenf du Cómité Ethique de
l'Hôpital Habib Thameur

Appendix 4: Example of informed consent for data sharing from the European Reference Network for Rare Diseases [18].

European Reference Networks

FORMULAIRE DE CONSENTEMENT DU PATIENT POUR LE PARTAGE DE DONNÉES

dans

LES RÉSEAUX DE RÉFÉRENCE EUROPÉENS POUR LES MALADIES RARES

pour

LES SOINS DES PATIENTS et LA CRÉATION DE REGISTRES DE MALADIES RARES

QUE SONT LES RÉSEAUX DE RÉFÉRENCE EUROPÉENS ET COMMENT PEUVENT-ILS M'AIDER ?

- Les réseaux de référence européens (ERN) sont des réseaux virtuels réunissant des prestataires de soins de santé qui s'occupent des maladies rares dans toute l'Europe. Ils ont été créés par la Directive 2014/24/EU relative à l'application des droits des patients en matière de soins de santé transfrontaliers.
- Ces réseaux ont été créés pour permettre aux prestataires de soins de santé de collaborer, afin d'aider les patients qui souffrent de maladies rares ou d'autres affections qui nécessitent un traitement hautement spécialisé.
- Avec votre accord et conformément aux lois européennes et nationales sur la protection des données, votre cas peut être soumis aux réseaux de référence, afin que les professionnels de la santé de ces réseaux aident votre médecin à poser un diagnostic et à mettre au point un programme de soins.
- Afin de permettre aux réseaux de référence de donner des conseils sur votre traitement, les données qui vous concernent et qui ont été collectées dans cet hôpital doivent être partagées avec des professionnels de la santé dans d'autres hôpitaux, parfois dans d'autres pays européens. Votre médecin peut vous en dire davantage sur les pays qui font partie des réseaux qui s'occupent de votre maladie.
- Les professionnels de la santé qui s'occupent habituellement de vous seront toujours responsables des soins qui vous seront apportés.
- Les données qui vous concernent ne seront pas partagées sans votre accord et, même si vous décidez de ne pas donner votre accord, les médecins continueront de s'occuper de vous le mieux possible.

QUELS SONT MES DROITS?

- Vous avez le droit de donner ou de retirer votre accord pour le partage de données avec le(s) réseau(x) de référence.
- Si vous acceptez aujourd'hui de partager vos données, vous pourrez toujours retirer votre accord plus tard. Votre médecin vous expliquera comment supprimer vos données des dossiers si vous le souhaitez. Il est possible que l'on ne puisse pas retirer des informations qui ont été utilisées pour vous soigner.
- Vous avez le droit d'être davantage informé sur les fins auxquelles vos données seront traitées et sur les personnes qui auront accès à ces données. Votre médecin vous dira qui peut vous aider si vous souhaitez plus d'informations
- Vous avez le droit de voir quelles informations vous concernant sont enregistrées et de demander des corrections si certaines informations ne sont pas correctes. Vous pourriez aussi avoir le droit de bloquer ou effacer vos données.
- L'hôpital dans lequel vos données sont collectées est responsable de ces données. Il devrait traiter vos demandes concernant vos données endéans les 30 jours.
- L'hôpital a le devoir de s'assurer que vos données sont traitées en toute sécurité et de vous avertir si la sécurité de vos données n'est pas respectée.
- Si vous vous inquiétez quant à la manière dont vos données sont traitées, vous pouvez contacter votre médecin traitant ou l'autorité compétente chargée de la protection des données.
- Votre hôpital évaluera tous les 15 ans le besoin de garder vos données dans le réseau de référence.

LES DONNÉES PARTAGÉES SONT RENDUES NON IDENTIFIABLES

- Si vous et vos médecins êtes d'accord pour demander le soutien de l'un ou de plusieurs réseaux de référence, ce formulaire de consentement autorisera l'hôpital à partager toutes les données de votre dossier médical qui pourraient aider les professionnels des réseaux à discuter des soins dont vous avez besoin.
- Votre nom et votre adresse ne seront pas communiqués.
- Les données partagées peuvent inclure des images médicales, des rapports de laboratoire ainsi que des données d'échantillons biologiques, ou encore des lettres et des rapports rédigés par d'autres médecins qui vous ont traités par le passé.
- Si vous décidez de consulter les réseaux de référence, vos données seront partagées grâce à un système électronique d'échange d'informations sécurisé appelé ERN Clinical Patient Management System.

QUID DES BASES DE DONNÉES/REGISTRES DE MALADIES RARES ?

- Pour améliorer les connaissances futures sur les maladies rares, les réseaux de référence européens dépendent beaucoup des bases de données visant la recherche et le développement des connaissances.
- Ces bases de données, aussi appelées registres, ne contiennent que des informations rendues non identifiables. Votre nom, votre date de naissance ou votre adresse ne sont PAS communiqués. Seules les informations sur votre état de santé y sont enregistrées.
- Afin de contribuer à ces registres, vous pouvez donner votre accord pour que les informations vous concernant y soient ajoutées. Si vous décidez de ne pas le donner, cela n'affectera en rien les soins qui vous sont apportés.

QUID DE LA RECHERCHE SUR LES MALADIES RARES?

- Vous pouvez aussi nous dire si vous souhaitez être contacté dans le cadre des projets de recherches pour lesquels vos données pourraient être utilisées.
- Si vous acceptez de partager vos données pour la recherche, nous vous contacterons pour que vous puissiez donner votre accord pour un projet de recherche précis.
- Vos données ne seront pas utilisées sans votre consentement pour un projet de recherche particulier.

CE FORMULAIRE DE CONSENTEMENT PEUT ÊTRE UTILISÉ POUR PARTAGER DES DONNÉES AVEC LE(S) RÉSEAU(X) DE RÉFÉRENCE SUIVANT(S)-

(À faire remplir par le prestataire de soins de santé qui signera au bas)

....

....

RENSEIGNEMENTS SUR LE PATIENT

Prénom : ....Nom de famille : ....

Date de naissance : ☐☐ ☐☐ ☐☐☐☐   Numéro d'identification : ☐☐☐☐☐☐☐☐☐☐☐☐☐

Veuillez cocher la case adéquate :

☐ Je suis le patient   ☐ Je suis le parent/tuteur du patient   ☐ J'ai une procuration

---

J'ACCEPTE que mes données rendues non identifiables soient partagées dans un ou plusieurs réseaux de référence européens pour mes **SOINS**.
Je comprends que mes données seront partagées avec des professionnels de la santé d'un ou plusieurs réseaux de référence pour qu'ils puissent collaborer en vue d'améliorer mes soins.

Signature   Date

....

JE REFUSE que mes données soient partagées dans un ou plusieurs réseaux de référence européens pour mes **SOINS**.
Je comprends que cela signifie qu'aucun réseau de référence ne peut être consulté pour contribuer à mes soins.

Signature   Date

....

J'ACCEPTE que mes données rendues non identifiables soient enregistrées dans un(e) ou plusieurs bases de données/ registres des réseaux de référence européens.

Signature   Date

....

JE REFUSE que mes données soient enregistrées dans une base de données/un registre des réseaux de référence européens.

Signature   Date

....

J'aimerais être contacté dans le cadre de recherches. Dans ce cas, je déciderai si j'accepte que mes données soient utilisées pour un projet spécifique ou non.

Signature   Date

....

Je ne veux pas être contacté pour que mes données soient utilisées pour la recherche.

Signature   Date

....

MÉDECIN TRAITANT ou PERSONNE AUTORISÉE À CONSTATER LE CONSENTEMENT

Nom ....   Fonction ....   Date ....

Appendix 5: Proposal for an informed consent form for sharing health data in the DMI

INFORMED CONSENT FORM FOR DATA SHARING

I, the undersigned ....(Name and surname of participant / Legal guardian),

☐ I freely and voluntarily accept (accept that the person under my guardianship)
Refuse (refuse that the person under my guardianship)

to share my health data (the health data of the person under my guardianship) on the computerized file with all health professionals, for the sole purpose of continuity and quality of care.
I acknowledge that I have received clear and intelligible information on the sharing of health data, its purpose and procedures, and that I have received answers to all my questions concerning this sharing.
I've had enough time to think things over and come to a decision.

I have also been informed of my right (the right of the person under my guardianship) to withdraw my consent (withdraw his/her consent) at any time if I deem it necessary or to define the data I may or may not share, and the healthcare professionals I wish to include in the sharing of my healthcare data without this causing me (him/her) any harm.
Done at ...., on .... Consent obtained by (Name and position) and electronic signature

Name of participant or legal guardian (Contact details: Last name and First name / telephone) and electronic signature

SHARED MEDICAL RECORDS IN THE ERA OF COMPUTERIZED MEDICAL RECORDS

Summary

Introduction :

The shared medical record (DMP) has only recently been introduced at the Habib Thameur Hospital (HHT). Its use raises ethical issues that have been little studied, especially in Tunisia. The aim of our work was to describe, through the experience of the HHT, the ethical issues raised by the DMP, notably information, confidentiality, and identification of the type of data and the objectives of data sharing (care, research...).

Methods :

This was a cross-sectional study (August 2023 - February 2024) which included professionals (qualified doctors and doctors in training) using the DMI at the HHT who had given their written informed consent to take part in the study. They filled in a questionnaire given or sent by email and including several pieces of information: general; concerning the computerized medical record, concerning the DMP and its ethical values as well as an open question to collect suggestions.

Results :

We included 86 participants (sex ratio M/F 0.43 and mean age 33.5 years [25-56 years]). Most were residents (n=29). All participants had prior knowledge of computer software.

participants. DMI use was standard practice in 85 cases.

The use of the DMI facilitated current practice for 80 participants, the negative points being mainly the lack of resources and training, and computer bugs. Forty-four participants thought that the medical record was unique to each hospital. Sharing health data optimized resource management, multidisciplinarity and overall patient care. Ethical reservations related mainly to the values of confidentiality and respect for medical secrecy (n=70), followed by communication/information (n=48). Sixty participants thought that informed consent was necessary before sharing health data (69%).

Conclusion:

At the end of our study, the participating doctors appear to be aware of the ethical dilemmas raised by the DMP. Awareness-raising, training of healthcare professionals, the development of common standards and the addition of informed consent will enable the DMP to be rolled out across the board in a way that respects the ethical concerns raised by the DMP.

essential ethical values.

Keywords: Bioethics, confidentiality, medical secrecy, computerized medical records, shared medical records.

SHARED MEDICAL RECORD IN THE ERA OF COMPUTERIZED MEDICAL RECORD

Abstract Background:

The shared medical record (SMR) was recently introduced at Habib Thameur Hospital (HHT). Its use raises ethical questions which have been little studied, especially in Tunisia. The aim of our study was to describe, through the experience of HHT, ethical considerations raised by the shared medical record, in particular information, confidentiality, and to identify objectives of data sharing (care, research, etc.).

Methods :

A cross-sectional study (August 2023 - February 2024) which including health professionals (graduate and training doctors) using the electronic medical record (EMR) in HHT who gave their written informed consent to participate in the study. They completed a questionnaire given or sent by email containing several informations: general; concerning the EMR, concerning the SMR and its ethical values as well as

Results :

Our study included 86 participants, mainly residents (n=29) with a sex ratio of 0.43 and a mean age of 33.5 years [25-56 years]. Previous use of computer software was noted among all participants. EMR was used in common practice in 85 cases. The use of EMR facilitated the daily practice of 80 participants; the negative points noted being mainly the lack of resources, training, and computer bugs. Forty-four participants believed that the medical record was unique per hospital. Sharing health data optimized resource management, multidisciplinarity and overall patient care. Ethical reservations expressed by the participants were mainly related to the values of confidentiality and respect for medical confidentiality (n=70), followed by communication/information (n=48). Sixty (69%) participants thought informed consent was necessary before sharing health data.

Conclusion:

Our study showed that the participating doctors seemed aware of the ethical dilemmas raised by the SMR. Raising awareness, training health professionals, developing common standards and adding informed consent will help to generalize the SMR while respecting essential ethical values.

Key-words: Bioethics, confidentiality, medical secrecy; computerized medical record, shared medical record

yes

I want morebooks!

Buy your books fast and straightforward online - at one of world's fastest growing online book stores! Environmentally sound due to Print-on-Demand technologies.

Buy your books online at
www.morebooks.shop

Kaufen Sie Ihre Bücher schnell und unkompliziert online – auf einer der am schnellsten wachsenden Buchhandelsplattformen weltweit! Dank Print-On-Demand umwelt- und ressourcenschonend produziert.

Bücher schneller online kaufen
www.morebooks.shop

Printed by Books on Demand GmbH, Norderstedt / Germany